Good Clinical Practice - A Guide to Archiving

2nd Edition
July 2014

Foreword

One of the fundamental requirements of the principles of Good Clinical Practice (GCP) is the need to ensure the integrity and secure retention of clinical trial documents for a period of time determined by legal, regulatory and business requirements. Essential Documents serve to demonstrate the compliance of the investigator, sponsor, and monitor with the standards of GCP and with all applicable regulatory requirements.

The first version of this guidance was published in 2007. The second edition has been produced to reflect changes in regulations since then, and to address feedback received on the first version.

The guidance and advice contained within this document are consistent with the requirements of EU Commission Clinical Trial Directive 2001/20/EC and supporting Directives, UK SI 1928:2006, US 21CFR312 and ICH GCP Guidelines.

Acknowledgements

The following members of the Scientific Archivists Group participated in the Working Party that produced this updated version of the guidance. Without their expertise and commitment, it could not have been produced.

Chris Jones, March 2014

Jennifer Barnwell
Liz Hooper
Russell Joyce
Mary Paul
Eldin Rammell
Cara Lax-Williams
Chris Jones

First Printing: 2007

Second Printing: 2014

ISBN 978-0-9557659-4-0

Scientific Archivists Group Limited

Registered Office Berkeley Townsend, Hunter House, 150 Hutton Road, Shenfield, Essex, CM15 8NL, UK

www.sagroup.org.uk

Contents

1 Scope

This document provides guidance to all organisations (Sponsors and Investigators) that are required to retain records in order to demonstrate compliance with Good Clinical Practice (GCP), and for all organisations that provide contract archive services for both electronic and physical records. Although aimed at GCP regulated organisations, the guidance and advice contained within this document might be of assistance to organisations that retain materials from other types of regulatory work, for example Good Manufacturing Practice (GMP).

When making decisions about the management of trial materials, organisations must ensure they evaluate applicable regulatory requirements as well as their own commercial and operational needs.

Certain aspects of archive construction and operation may be affected by local or national building regulations or health & safety legislation: guidance on these, and other such aspects, is outside the scope of this document.

2 Definition of Terms and Abbreviations

Archive: The physical or electronic facility designated for the secure retention and maintenance of archived materials, including the operation of that facility under the control of an archivist

Archivist: The individual responsible for the day to day operation and management of the archive in accordance with organisational policies, standard operating procedures and GCP. This responsibility should be documented, for instance in job descriptions or organisation charts.

Computer System Validation: The ongoing process of establishing documented evidence that provides a high degree of assurance that a computerised system will consistently perform according to its predetermined requirements and quality attributes. This includes procedures, requirements and specifications, testing, and change control.

Contract Archive Facility: An organisation contracted by the Sponsor, Investigator or other institution to retain trial material under the conditions required by GCP and as defined in this document.

Contract Research Organisation (CRO): A person or an organisation (commercial, academic or other) contracted to perform one or more trial related duty or function by the Sponsor.

Data Management: The collection, maintenance and handling of data generated during the conduct of a clinical trial.

Essential Documents: Documents that individually and collectively permit evaluation of the conduct of a trial and the quality of the data produced. (See ICH GCP Section 8 for identification of the minimum list of essential documents and the responsibilities for their retention).

Ethics Committees: Group of suitably qualified and experienced people who review and evaluate the science, medical aspects and ethics of a proposed clinical trial.

Investigator Site: Institution or facility where the clinical trial, or part of a trial is conducted.

Investigator Site File: A file at the Investigator site that contains all the essential documents relating to a clinical trial, before the trial commences, during trial conduct and after the trial.

Legal Hold (Preservation Hold): A legal directive requiring that records related to a specified product or issue must be preserved and must not be destroyed until further notice. When a legal hold is issued, it takes precedence over the normal record retention timeframes and will require preservation of the affected records, even if a record would otherwise have to be discarded under the applicable record retention schedule.

Material(s): A collective term given to all items required to be retained for regulatory purposes. This includes, but is not limited to all “essential documents”, source data, specimens, and non trial-specific records. This includes records maintained in electronic form.

Obsolescence: The process by which computer hardware, software, operating system or storage media becomes unusable by virtue of the fact that there is no longer a system readily available that is capable of using the hardware or storage media or running the software or operating system.

Pharmacovigilance: The science and activities relating to the detection, assessment, understanding and prevention of adverse effects or any other drug-related problem.

Quality Audit: A defined system, including personnel, which is independent of the conduct of a clinical trial and designed to assure compliance with regulations, contracts and organisation standard operating procedures (SOPs).

Record Retention Schedule: A list of material types with defined retention periods.

Retention Period: The length of time for which materials should be retained before destruction.

Sponsor: The individual or organisation that takes responsibility for the initiation, management and/or financing of a clinical trial.

Storage Media: The different materials on which information may be recorded. Examples include paper, photographic film, magnetic media, microforms and optical devices.

Trial Master File (TMF): A file or collections of files that contain all the essential documents relating to a clinical trial, before the trial commences, during trial conduct and after the trial at both the Sponsor and the Investigator site.

3 Roles and Responsibilities

3.1 Sponsor

Prior to initiating a trial the Sponsor should define, establish and allocate all trial related duties and functions.

In relation to material retention, the Sponsor should:

- Agree the format in which all records will be maintained throughout their lifecycle.
- Ensure that the materials required to support their studies are retained and maintained in conditions that ensure its integrity and continued access.
- Establish a retention schedule for all material created.
- Maintain contact with the organisation that archives its materials.
- Ensure procedures are established for making materials available for inspection e.g. by regulatory authority(ies).

Where trial activities are outsourced to a third party such as a Contract Research Organisation or Electronic Data Capture vendor, particular care should be taken to ensure that the third party understands their responsibilities with respect to the creation and management of data, documents and records. It is recommended that records management and archiving responsibilities are documented and agreed between the two parties ideally prior to commencement of the trial and, where appropriate, included in the contractual agreement. Archiving arrangements should include the timing, process and format for the transfer of materials from the third party back to the Sponsor and any special arrangements made regarding the custody of specialist materials (i.e. retained samples) post trial completion.

In relation to material retained by the Investigator, the Sponsor should:

- Assess the Investigator's capability to retain trial material in appropriate and suitable conditions.
- Inform the Investigator in writing of the need for material retention (as part of the contract) including timeframes.

The Sponsor may provide financial assistance if Investigator sites are incapable of retaining material, whilst ensuring the materials remain the responsibility of the Investigator. The Sponsor must not have direct access to such material.

3.2 Investigator(s)

The Investigator should recognise the importance of good records management practices and comply with the requirements of the Sponsor (as per contractual agreement), provided these meet or exceed appropriate regulatory requirements and local operating procedures.

In relation to materials at the Investigator site, the Investigator should:

- Retain responsibility and accountability for the trial materials throughout their lifecycle.
- Take appropriate measures to minimise risk of loss, damage or destruction of materials.
- Inform the Sponsor in writing of any changes in contact details/ personnel who have control of the archived trial material.
- In the event the Investigator is unable to provide suitable storage for the trial materials, a contract archive facility may be used. In such an event the Sponsor should be informed of the location of the material. Such site material should not be returned to the Sponsor.
- Ensure procedures are established for making materials available for inspection e.g. by regulatory authorities.

- If not the Investigator, an individual should be identified by the Investigator and appointed as the Archivist to fulfil the responsibilities defined in this guidance.

3.3 Ethics Committees (EC) & Institutional Review Boards (IRB)

The purpose of EC/IRB is to safeguard the rights, dignity and welfare of subjects participating in research. The EC/IRB is entirely independent of the researcher (Investigator) and the organisations funding and hosting the research (Sponsor, CRO, etc).

The EC/IRB should establish, document in writing and follow its procedures.

ICH E6 Section 3.4 states the EC/IRB should retain all relevant records for a period of at least 3 years after completion of a trial and make them available upon request to regulatory authorities.

The principles defined by this guidance document should equally be applied by such EC/IRB to the records retained.

3.4 Third Party Service Providers/CROs

A Sponsor or Investigator may subcontract trial related activities to a third party service provider.

Examples of the type of function that may be subcontracted include: laboratory investigations, site monitoring, data management, statistics, report writing and archiving of trial materials.

This division of accountabilities, including the maintenance of materials created as a result, must be clearly documented in the contractual agreement. The Sponsor retains overall responsibility for the conduct of the trial and for the quality of all trial materials. The Sponsor or Investigator should assess the service provider's capability to maintain trial material in appropriate and suitable conditions during the conduct of the trial. If archiving is delegated to a service provider, the service provider's capability to maintain trial materials for the agreed period of time must similarly be assessed.

3.5 Archivist

The Archivist is responsible for the day-to-day operation and management of the archive in accordance with documented company policies, operating procedures, GCP, the guidance contained within this document and any contractual obligations. The Archivist should be formally identified. In larger organisations, there may be a need for several staff to support the archive function.

The role of Archivist does not have to be a dedicated role; it can be combined with other roles, dependant on the available resource and workload within the organisation. It should be noted that in the case of electronic materials, archiving is often the responsibility of resources within the I.T. function. This is acceptable provided the regulatory requirements pertaining to the archive and the archivist role described in this document are satisfied.

All Archivists and archive staff should undergo a level of training appropriate to their role and be able to demonstrate an appropriate level of competence in the performance of their role. The training should be documented and reviewed on a regular basis.

Material cannot be considered to have been archived unless it comes under the control of an Archivist.

Archivists are responsible for the following:

- Ensuring that the archive is operated in accordance with guidance defined in this document and any specific archiving documentation, standard operating procedures and work instructions for their organisation.
- Monitoring the conditions within the archive and addressing any issues arising.
- Indexing, identifying and managing materials deposited into the archive.
- Controlling access to the archives and archived material in accordance with defined procedures.
- Managing, tracking and documenting material retrieval, loan, transfer and destruction.

Contract archive facilities may be utilised in the retention of trial materials (Section 14).

4 Control of Active Documentation

The duration of clinical trials from trial initiation to trial close-out may be just a few days for a Phase I trial through to several years for large Phase III trials. The timing and procedures for the transfer of trial materials to the archive is dictated by the Sponsor or Investigator site policy and must be agreed and documented. Prior to archiving, materials should be managed as active or semi-active records. To assist in this process consideration should be given to the following:

- Assigning an individual to be responsible for the active trial materials until they are deposited and accepted into the archives.
- Maintaining the materials in a secure environment, where they are accessible to appropriate staff only.
- Maintaining the materials in a location where they are at low risk from loss or damage due to environmental factors, such as fluctuation of temperature and humidity, flood, pest control, etc.
- Describing processes for managing active records in appropriate trial management policies or procedures.

5 The Archive

The following section describes archives for the storage of physical materials. However, the principles described should also be applied to the storage of electronic materials. Further specific issues relevant to electronic records are described in "A Guide to Archiving of Electronic Records".

5.1 Archive Design

An archive should be suitably designed and constructed to accommodate the materials to be archived and to ensure their integrity once placed within the archive. A building, room, fire-resistant safe, filing cabinet or other appropriate storage can be designated as an archive provided it meets the requirements as defined in this document.

The following factors should be considered as part of the archive designation:

- A secure location to prevent unauthorised access to the retained materials, e.g. by the use of locks or electronic entry systems.
- Construction designed to withstand the elements of weather. Consideration may need to be given to specific local conditions such as a risk of flooding.
- Running water pipes in or near the archive area should be avoided, as there is a risk of leakage and subsequent water damage.
- It is recommended that an automated fire detection system is installed, and consideration may also be given to installing an automated fire suppression system if feasible.
- The use of naked flames, or other open heat sources, within or around the archives should be prohibited. The use of electrical appliances should be minimised.

- A risk assessment relating to the entry of pests (e.g. rodents, insects) into the archive should be undertaken. If appropriate a monitoring program should be implemented.
- A location that is assessed to be at minimum risk of disruption from potential adverse events in the environment, at local businesses, or from transport infrastructures for example.

5.2 Archive Storage Conditions

Archive storage conditions must not adversely affect the physical integrity of archived materials or the integrity of the information.

Fluctuations of temperature and humidity may contribute to the deterioration of materials; it is therefore important to ensure consistent environmental conditions. Acceptable storage conditions for the different types of archived materials should be defined.

The following factors should be considered:

- The installation of systems or processes to monitor and control environmental conditions may be necessary. This should be supported by procedures for monitoring, defining and addressing out of specification measurements.
- In situations where there is a significant risk of temperature and humidity variation, or prolonged exposure to conditions outside of acceptable storage conditions, it will be necessary to exercise more frequent checks of stored materials to confirm their integrity.
- Special storage conditions may be required for particular materials, e.g. materials that need to be stored frozen, refrigerated, etc., or free from dust or magnetic interference (as in the case of electronic media).

Guidance on storage conditions may be found in the reference documents in section 17 of this document.

5.3 Disaster Recovery and Business Continuity

A disaster recovery plan describing the steps to be taken in the event of damage to or inaccessibility to the archive or archived materials should be in place. The types of event to be considered in the plan should be determined by an assessment of the risks to which the archive or its operation may be exposed. These risks are likely to be most dependent on the archive location and environment. Examples of the risks to be considered should include but not be limited to fire, flood, pest infestation and structural damage, e.g. subsidence, forced entry of the archive by unauthorised persons.

It is recommended that a disaster recovery plan/procedure covering an archive includes:

- Plan or outline of building, including location of different collections of archived materials.
- Process for recovery and/or restoration of lost or damaged materials and re-establishing the security of the archive.
- Names and telephone numbers of appropriate personnel who can be contacted outside of normal working hours.
- Whereabouts of equipment to support recovery and restoration of archive materials (e.g. dehumidifying equipment, wrapping materials etc).
- Possible roles of the emergency services (it may be useful to liaise with them so they understand special concern of archives and libraries for possible water damage).
- Define who may access the archive in an emergency.

- Identify that access to the archive by employees will be denied while there is a risk to health and safety.
- The records that should be made to document the disaster and recovery effort.

It may be useful to establish contracts with appropriate specialist organisations that can provide a service to recover and/or restore damaged archived materials. It is recommended that disaster recovery plans are tested periodically.

A business continuity plan identifies risks to the operation of the archive and measures to reduce or mitigate those risks. Often this term is used interchangeably with disaster recovery. However there are many risks that are not “disaster” related. These can include:

- unavailability of archive indexing system for a period of time affecting retrieval or depositing of new records
- archive staff sickness
- issues with a third party supplier that the archiving service is dependent upon (i.e. commercial records storage contractor)

The plan should identify the likelihood of the risks occurring, and their potential impact – which should then help determine, what, if any mitigations need to be established.
Disaster recovery and business continuity plans should be reviewed regularly and tested to confirm they are fit for purpose.

6 Operation of the Archive

6.1 Transfer of trial materials into the archive

6.1.1 Submission of materials for archiving

There should be documented procedures in place which define the processes for archiving of trial materials, including checks for completeness, timeframes for submission to archive, delegation of responsibility for submission of material to the archive etc.

It is recommended that all trial materials, not already archived, are archived promptly on completion of the trial –or as immediately as possible thereafter- as defined by the Sponsor. The maximum time period for transfer of materials to the archive following trial completion should be agreed and documented. This will ensure that any remedial action required can be exercised promptly and effectively. Where the trial materials are being generated by a CRO, it is recommended that the time period for transfer to the Sponsor allows sufficient time afterwards for quality checks by the Sponsor to be made, before final transfer of the materials to the Sponsor archive within agreed timeframes.

6.1.2 Receipt of materials into the archive

Archive staff should verify that the materials received correspond with the information provided in support of the materials and that any deficiencies are dealt with promptly. It is recommended that the Sponsor and Investigator have a standard TMF contents list against which they can check completeness of the files; this may be based upon the TMF Reference Model. There should be a documented acceptance of the trial material by the archive staff to provide a traceable chain of custody record.

6.1.3 Indexing of archived materials

Archived materials should be indexed by appropriately trained staff in a manner to facilitate ease of retrieval. The indices of archived materials can be electronic or paper based although the former provides a greater degree of flexibility in terms of ease of retrieval and the maintenance of currency of any archive collection. The indices should contain sufficient detail to support the retrieval, review loan management, content management, and disposal of the archived materials in the future.

6.1.4 Filing of archived materials

Archived materials should be filed to ensure their continuing integrity and to facilitate ease of retrieval. Containers should be stored safely to prevent damage to the contents and minimise risk to archive staff who handle them.

6.1.5 Access to the Archive

Access to an archive should be controlled and restricted to the Archivist, archiving staff and other named individuals. A regular (i.e. annual) review of the access list should be undertaken and documented. A record of visits by anyone not on the list of named individuals should be retained. This record should include the identity of the visitor, the reason for the visit and the date/time of the visit.

6.2 Non-trial specific materials

Non-trial specific materials, such as SOPs, staff training records, organisation charts, computer system validation documentation, contractual documentation, etc., should be maintained according to organisational policies.

6.3 Trial Samples and Specimens

Only samples that afford re-evaluation or support reconstruction of the trial need to be retained. These include pathology samples and slides, blood slides and tissue samples. Such samples as urine, serum and plasma used as part of the safety assessment process do not need to be retained. An organisation may have non-trial related reasons and permission to retain material beyond the end of the trial (e.g. tissue banks).

The retained samples should be archived under conditions that minimise their deterioration.

6.4 Retrieval of Archived Materials

Materials that have been archived may, on occasion, need to be retrieved, e.g. for regulatory inspection. Where retrievals from the archive are permitted, these should be managed according to approved documented procedures.

These procedures should at a minimum include:

- Details of management controls related to access to archived materials (e.g. which materials can be accessed by whom, any authorisation needed to support removal of materials from the archive etc).
- The reasons for which materials may be removed from the archive (often referred to as loans).
- The timeframe within which the removed (loaned) material should be returned to the archive.
- How the Archivist identifies any materials that have not been returned within the specified timeframe.
- Mechanisms to ensure the return of these materials by the person to whom they had been loaned.

On return to the archive, the person to whom the materials have been loaned should identify changes or additions that have been made, in accordance with appropriate procedures.

Records of all material loans and subsequent returns should be retained by the archive. In the case of an electronic archive, if security controls prevent materials from being removed (e.g. only a copy is downloaded for viewing purposes) or prevent changes from being made, it is not considered necessary to retain a record of accesses to those archived materials.

7 Transfers of Archived Materials

On occasion it may be necessary to transfer archived materials from one location to another, e.g. to support transfer of ownership, relocation of archive facilities, etc. In the case of electronic materials, transfer between storage systems may be necessary to prevent obsolescence or deterioration of storage media or format.

The archived materials to be transferred should be clearly described in appropriate chain of custody documentation. This should include details of the materials, the contact details/address of the receiving facility, and the means of transfer between locations.

The transportation of the material, and associated paperwork, between the two locations should be undertaken in such a way as to minimise the risk of loss or damage of the materials.

The recipient of the transferred materials should check that they correspond with the associated chain of custody documentation as, once accepted, they become responsible for ensuring that the data is maintained and preserved appropriately. In the case of electronic materials, the transfer process must be validated. Copies of the chain of custody/validation documentation should be retained by all parties involved in the transfer.

8 Investigator Site Records Management

Throughout the conduct of the trial, the Investigator is required to manage trial materials to protect their confidentiality and integrity, in line with their contractual obligations to the sponsor. Once the trial is complete, the Investigator is required to archive trial materials to ensure their security, confidentiality and integrity, in line with their contractual obligations to the Sponsor, until written confirmation is received from the Sponsor that they can be destroyed.

Consideration should be given to the following:

- Security: the materials should be stored securely to prevent unauthorised access.
- Storage conditions: the materials should be stored to minimise risk due to fire, flood, pest, etc.
- Confidentiality: the materials supporting different Sponsors studies should be separated to avoid cross contamination.
- Sponsor access (during trial): the contract should define how, and when, stored material can be accessed by representatives of the Sponsor.
- Sponsor access (after trial completion): the materials should be stored to ensure that there is no risk of unauthorised access by the Sponsor. It is acceptable for the Sponsor to provide financial assistance if a contract archive facility is to be used.
- Retrieval: the materials should be catalogued in order to ensure they can be made readily available upon request.
- Medical records should be labelled to avoid accidental destruction: e.g. place a 'Do Not Destroy' label on the records of subjects who consented to be in a clinical trial and mark the archive inventory accordingly.

It is acceptable for materials generated electronically to be retained electronically if agreed with the Sponsor and Investigator. However, it is important that the file format selected and storage medium selected are suitable for the storage of materials for a minimum period of 15 years. Some storage media, for example CDs/DVDs and flash drives, have a relatively short life expectancy, are prone to accidental damage, more likely to be lost and consequently are not considered suitable for long-term storage of data.

Following expiry of the retention period and once confirmation of permission has been received from the Sponsor the materials should be destroyed according to the Investigator's site policy and any Sponsor requirements. A certificate of destruction may be produced and retained by the site.

9 Electronic Records

An electronic record is defined as information recorded in electronic form that requires a computerised system to access or process. The risk of obsolescence (to media, software or file format) of electronic records presents special challenges for their long term retention. These challenges need to be considered in the design of systems used to support clinical trials. When defining how electronic records will be maintained, consideration should be given to the following risks to readability throughout the retention period:

- Potential hardware obsolescence.
- Potential operating system obsolescence.
- Potential software obsolescence.
- Potential file format obsolescence.
- Storage media deterioration/obsolescence.

To counter the above risks, and ensure that essential information remains complete and retrievable throughout the retention period, a migration strategy is essential.
The strategy must ensure the completeness, accuracy, reliability and trustworthiness of the data transferred. The migration procedures used must be validated and the validation documented.

Electronic records from completed clinical trials should be maintained in a designated archive. Procedures must be in place to ensure they are managed appropriately throughout their lifecycle.

Consideration should be given to the following:

- Media: the shelf-life of the storage media. A review schedule should be set up to read / test records held on media on a regular basis, based on a risk assessment.
- Physical security: access to hardware and storage media should be limited to authorised staff.

- Environmental conditions: storage media requires an environment that has minimal fluctuations in temperature and humidity.
- Access control: there should be security measures in place to ensure that only authorised personnel can access the electronic records, and to enforce restrictions on access levels.
- Record integrity: controls must be implemented to ensure that archived records cannot be modified. Records may only be deleted with appropriate authorisation.
- Accessibility: electronic records must be available upon request.
- Disaster Recovery: disaster recovery copies of all data kept on electronic media should be made and kept in a separate location.

Comprehensive guidance on archiving of electronic records is available in the SAG publication “A Guide to Archiving of Electronic Records”.

10 Retention Period of Materials

The retention period set for trial materials should take account of all legal, regulatory, operational, national, historical or fiscal requirements that apply to the organisation.
It should be noted that although ICH guidelines and EU directives provide guidance on the regulatory expectations, it is important that the Sponsor and Investigator take account of other requirements. These requirements may conflict with each other; for example, the requirement to only retain personal data for as long as is necessary (Data Protection legislation) and a business requirement to maintain protection against product litigation. The retention period for trial materials must therefore be decided on a risk-based approach (the level of risk will differ from organisation to organisation).

10.1 Applicable Regulations

European Directive 2005/28/EC Article 17 states:

“The sponsor and investigator shall retain the essential documents relating to a clinical trial for at least five years after its completion.

They shall retain the documents for a longer period, where so required by other applicable requirements or by an agreement between the sponsor and the investigator.”

European Directive 2003/63/EC, Annex 1, Chapter 5 states:

“Marketing authorisation holders must arrange for essential clinical trial documents (including case report forms) other than subject’s medical files, to be kept by the owners of the data:

- for at least 15 years after completion or discontinuation of the trial,
- or for at least two years after the granting of the last marketing authorisation in the European Community and

when there are no pending or contemplated marketing applications in the European Community,

- or for at least two years after formal discontinuation of clinical development of the investigational product.

Subject's medical files should be retained in accordance with applicable legislation and in accordance with the maximum period of time permitted by the hospital, institution or private practice.

The documents can be retained for a longer period, however, if required by the applicable regulatory requirements or by agreement with the sponsor. It is the responsibility of the sponsor to inform the hospital, institution or practice as to when these documents no longer need to be retained.

The sponsor or other owner of the data shall retain all other documentation pertaining to the trial as long as the product is authorised. This documentation shall include: the protocol including the rationale, objectives and statistical design and methodology of the trial, with conditions under which it is performed and managed, and details of the investigational product, the reference medicinal product and/or the placebo used; standard operating procedures; all written opinions on the protocol and procedures; the investigator's brochure; case report forms on each trial subject; final
report; audit certificate(s), if available. The final report shall be retained by the sponsor or subsequent owner, for five years after the medicinal product is no longer authorised."

ICH E6 Section 5.5.11 states:

"The sponsor specific essential documents should be retained until at least 2 years after the last approval of a marketing application in an ICH region and until there are no pending or contemplated marketing applications in an ICH region or at least two years have elapsed since the formal discontinuation of clinical development of the investigational product. These documents should be retained for a longer period however if required by the applicable regulatory requirement(s) or if needed by the sponsor."

Detailed guidelines on good clinical practice specific to advanced therapy medicinal products (December 2009) from EudraLex vol 10 states:

"The sponsor of the trial, the tissue establishments/ procurement organization, the animal facility, the manufacturer and the investigator/institution where the ATIMP is used, should keep their parts of the traceability records for a minimum of 30 years after the expiry date of the product, or longer if required by the terms of the clinical trial authorization or by the agreement with the sponsor."

US regulations FDA 21 CFR 312.57 states:
"A sponsor shall retain the records and reports required by this part for 2 years after a marketing application is approved for the drug; or, if an application is not approved for the drug, until 2 years after shipment and delivery of the drug for investigational use is discontinued and FDA has been so notified."

10.2 Sponsor Responsibility

The Sponsor should define the conditions under which materials should be retained, and their retention period.

The Sponsor requirements for retention of materials at the Investigator site should be documented in the contract. Inform the Investigator site in writing when trial materials are no longer required to be retained. The Sponsor should retain records of their attempts to contact the Investigator.

10.3 Investigator Trial Sites

The materials to be retained, their retention periods and the responsibility for the Investigator to manage the materials in an appropriate manner should be agreed and defined by the Sponsor prior to trial start and formally documented in the contract.

The retention period for patient records and some other source documentation are defined by hospital, institution or private practice policy. If that retention period is shorter than that required for the trial, these records should be retained until the Sponsor gives permission for their destruction. It is recommended that trial materials are clearly identified as such to minimise the risk of premature destruction.

11 Destruction of Material

The Investigator and Sponsor should take measures to prevent accidental or premature destruction of trial materials.

There should be defined procedures for managing the destruction of trial materials that include

- Processes for identifying materials that have reached the end of their retention period.
- Approvals needed for the destruction to occur. Such approvals should be made by appropriately qualified individuals.
- The method by which the destruction will be carried out whilst ensuring the materials confidentiality is maintained.
- Records of destruction that should be retained.

For Investigator site materials, approval from the Sponsor and the Investigator should be obtained before destruction occurs. In exceptional circumstances, in the event that either party is not available to sanction the destruction, the other party can authorise the destruction if the retention criteria has been fulfilled. Records of attempts to contact the other party should be retained.

12 Pharmacovigilance records

The management of pharmacovigilance records generated during clinical trials is guided by a number of regulations in addition in GCP.

The retention period of Pharmacovigilance documentation generated in the conduct of a clinical trial is different than that defined for clinical trial documentation within the scope of ICH GCP. The scope of pharmacovigilance includes marketed products as well as investigational products. Current EU regulatory guidance recommends indefinite retention of Sponsor Pharmacovigilance documentation for all products.

The same general principles should be taken into account in relation to the storage and archiving of pharmacovigilance documentation, as for any other material from a clinical trial as defined by this document.

However special consideration should be given to:

- Access controls: due to the highly confidential nature of the materials.
- Retrieval: as speed of response to queries is of paramount importance.

13 Quality Audit

The Sponsor is responsible for ensuring the implementation of a quality audit process for all aspects of the conduct of clinical studies as part of the quality management system. This must include facilities used for the retention of trial related material. Quality audits must be conducted according to an audit plan to ensure material from the clinical trial is being retained and maintained in accordance with GCP and this guideline.

Quality audits must be conducted by appropriately trained and competent staff.
Written reports of such audits should be produced in a timely manner and issued to the Sponsor. Such audit reports with documented corrective actions should be retained.

The frequency of quality audits will be influenced by the type of archive arrangements in place, the number of trials being conducted and an assessment of potential risks or breaches of contracts that potentially impact the integrity of archived materials.

14 Contract Archive Services

The use of contract archive facilities by Sponsors, Investigators or third party service providers to hold clinical trial materials, including electronic materials, is permissible. The guidance on management of archive facilities contained within this document equally applies to all contract facilities used to store trial materials.

There should be a formal, documented, agreement between the contract archive and the Sponsor or Investigator that details the level and conditions of service to be provided.

This agreement should include:

- The specific location of archive facilities to be used.
- Process for authorising use of alternative archive facilities.
- Transfer of material to and from the archive.
- Chain of custody of materials.
- Access rights to stored material.
- Storage conditions within the archive facility and required monitoring.
- Period of storage.
- Method of retrieval/access.
- Method of return/disposal including authorisation requirements.
- Service level agreements in relation to the services provided.

The contract archive organisation should have documented procedures for its operations. In addition, records of recruitment and staff training should be maintained.

15 Closure of an Archive

If an Investigator site closes down, the fate of any archived materials held at that facility needs to be determined. Materials should be retained in a secure manner to ensure their future integrity.

When the archive that is closing down is part of a larger organisation, there is the possibility that retained materials could be transferred to other archives within that organisation, following transfer procedures such as those noted in section 7.

If the archive that is closing down is owned by a third party, in the first instance they should contact the Sponsors of the clinical trial and ask for advice on the transfer or disposal of any retained materials. Possible scenarios are:

- The Sponsor may request that materials are returned to them. To ensure the validity of the trial data, it might also be necessary for the Sponsor to obtain copies of relevant non-trial specific material.
- The Sponsor may request that their materials be transferred to another storage location specified by them. Again, it might be necessary for them to obtain copies of relevant non-trial specific material that would be necessary to support the trial data.
- The Sponsor may instruct the third party to destroy some, or all of their materials.

In each case, the Sponsor should ensure that there should be complete and accurate records of the final disposition of all data and records.

16 Legal

In the event of an Institution/Sponsor take over, merger, close-down, or product acquisition, line management and the legal department should consult with the Archivist for guidance as appropriate. The property rights for archived records should be defined by Institution/Sponsor management and not by the Archivist.

17 Other Useful Reading

British Standards Institute Published Document: PD5454:2012 *Guide For the Storage and Exhibition of Archive Materials*

European Commission: *European Union, Directive 2001/20/EC (Clinical Trials Directive).*

European Commission: European Union, Directive 2003/63/EC (Clinical Trials Amendment)

European Commission: *European Union, Directive 2005/28/EC (Good Clinical Practice Directive).*

European Forum for Good Clinical Practice (EFGCP): *Guidelines for Retention of Clinical Trial Records at investigator Study Sites*. Update Paper Version 25 Mar 1995.

Food and Drug Administration: *Guidance for Industry: Title 21 Code of Federal Regulations Part 11 (21 CFR Part 11) Electronic Records; Electronic Signatures.*

Food and Drug Administration: *Guidance for Industry: Title 21 Code of Federal Regulations Part 312 (21 CFR Part 312)Investigational New Drug Application.*

Good Clinical Practice Records Managers Association "Guide to Selecting a Commercial Records Storage Provider" (http://gcp-rma.org/publications/)

Hutchinson. D: The *Trial Investigators GCP Handbook, a practical guide to ICH requirements*. 1997 (Brookwood booklets).

Hutchinson. D: *Which Documents, why? A guide to essential clinical trial documentation for Investigators*. 1997 (Brookwood Medical Publications).

International Organisation for Standardisation (ISO): ISO 11799: 2003 *Information and documentation—Document storage requirements for archive and library materials.*

Medicines and Healthcare products Regulatory Agency: *FAQs for Trial Master Files (TMF) and Archiving 2011.*

Medicines and Healthcare products Regulatory Agency: *Good Clinical Practice Guide,* 2012

National Archives: *Electronic records management strategy documentation and procedures from the National Archives* www.nationalarchives.gov.uk

ICH GCP Guidelines for GCP: 1996

Statutory Instruments 2006 No. 1928: *The Medicines for Human Use (Clinical Trials) Amendment Regulations 2006.*

APPENDIX 1 - Recommended archive standards

The Working Party recognises the challenges of implementing this guidance in all clinical trial situations.

The following information

- provides guidance on the recommended standard for the design, management and operation of a GCP regulatory archive;
- outlines the minimal standards for the operation of an archive, including assessment of associated risks; and
- provides guidance on the storage of clinical trial records outside of an archive.

PEOPLE:

Criteria	Responsible person
Recommended archive standards	Appointed Archivist with responsibility for and control of the archive.
Minimum archive standards	Appointed Archivist with responsibility for and control of the archive.
Risk of using minimum archives standard	N/A
Storage standards for ongoing trial materials	Nominated person with responsibility for storage of materials.

Criteria	Training
Recommended archive standards	Training for all staff working in archive covering GCP regulations, records management principles and archive procedures. Regular refreshers provided for all appropriate staff.
Minimum archive standards	Training for all staff covering archiving procedures. Training provided when procedure changes.
Risk of using minimum archives standard	Lack of awareness of up to date knowledge of GCP and risk of procedures being non-compliant.
Storage standards for ongoing trial materials	Training in good records management.

Criteria	Training file
Recommended archive standards	Training file including Job description, CV and details of relevant training received.
Minimum archive standards	Training file containing details of training received.
Risk of using minimum archives standard	N/A
Storage standards for ongoing trial materials	Training file containing details of training received.

FACILITIES:

Criteria	Archive design and operation
Recommended archive standards	Suitably designed, equipped and operated as a regulated archive.
Minimum archive standards	Adapted for use as suitable long term storage.
Risk of using minimum archives standard	Increased need to monitor facility and operation of facility.
Storage standards for ongoing trial materials	N/A

Criteria	Location
Recommended archive standards	Geographic location to minimise risk.
Minimum archive standards	Minimise risk of flooding.
Risk of using minimum archives standard	N/A
Storage standards for ongoing trial materials	Minimise risk of flooding.

Criteria	Security - General
Recommended archive standards	Site CCTV, 24 hr guards and regular patrols, intruder alarms and perimeter fencing.
Minimum archive standards	Locked building with intruder alarms.
Risk of using minimum archives standard	Greater risk to archived materials from unauthorised access
Storage standards for ongoing trial materials	Locked containers or cupboard. Limited access to a key.

Criteria	Security – Access control
Recommended archive standards	Restricted access to archive and access monitoring.
Minimum archive standards	Restricted access.
Risk of using minimum archives standard	N/A
Storage standards for ongoing trial materials	Access to materials is controlled by nominated person.

Criteria	Security – Vetting
Recommended archive standards	All staff who work in the archive to have been security vetted.
Minimum archive standards	References checked
Risk of using minimum archives standard	N/A
Storage standards for ongoing trial materials	References checked.

Criteria	Security – Visitor access
Recommended archive standards	Non-archive staff never left unaccompanied in the archive.
Minimum archive standards	N/A
Risk of using minimum archives standard	N/A
Storage standards for ongoing trial materials	N/A

Criteria	Fire Detection
Recommended archive standards	Fire detection system and automated fire suppression systems. Fire alarms linked to security.
Minimum archive standards	Fire alarms and automated alert.
Risk of using minimum archives standard	Increased risk of damage to records due to delay in response.
Storage standards for ongoing trial materials	Smoke detection system and extinguishers.

Criteria	Fire Detection – Testing
Recommended archive standards	Routine testing /maintenance of alarms, alerting and suppression systems. Regular tests of fire drills.
Minimum archive standards	Routine testing /maintenance of alarms and alerting systems.
Risk of using minimum archives standard	N/A
Storage standards for ongoing trial materials	Regular checking of smoke detection system and extinguishers.

Criteria	Temperature and Humidity Monitoring
Recommended archive standards	Automated monitoring and recording against defined ranges. Procedure for addressing out of specification values.
Minimum archive standards	Monitoring and recording against defined ranges.
Risk of using minimum archives standard	Resources required to do manual checking in all storage locations.
Storage standards for ongoing trial materials	Check that records are not adversely affected by their environment.

Criteria	Pests
Recommended archive standards	Stringent pest prevention program. Frequent monitoring and records retained.
Minimum archive standards	A pest prevention and monitoring program.
Risk of using minimum archives standard	N/A
Storage standards for ongoing trial materials	Respond to evidence of pests.

Criteria	Disaster recovery
Recommended archive standards	Extensive disaster recovery plan, associated training and regular review and testing program.
Minimum archive standards	Written disaster recovery plan (can be part of organisations plans).
Risk of using minimum archives standard	N/A
Storage standards for ongoing trial materials	Should be covered in organisations business continuity plan.

Criteria	SLAs contracts with third party
Recommended archive standards	Contract in place, including extensive Service Level Agreement (SLA) and routine monitoring.
Minimum archive standards	A signed contract with minimal SLA information.
Risk of using minimum archives standard	Not all aspects of service well defined, hence unclear issue resolution.
Storage standards for ongoing trial materials	A signed contract.

PROCEDURES:

Criteria	Depositing
Recommended archive standards	Clearly defined and documented process for transfer of materials into the archive.
Minimum archive standards	Records of materials received.
Risk of using minimum archives standard	Discrepancies between sender and receiver cannot be readily addressed.
Storage standards for ongoing trial materials	N/A

Criteria	Index of archive holdings
Recommended archive standards	Comprehensive, robust index. If electronic system is used, validation necessary.
Minimum archive standards	Reliable index of filed materials.
Risk of using minimum archives standard	Slower response time for retrievals.
Storage standards for ongoing trial materials	Simple filing structure to aid retrieval.

Criteria	Access to archived records
Recommended archive standards	Access to archived records is strictly controlled via defined procedures.
Minimum archive standards	Access to archived records is monitored.
Risk of using minimum archives standard	Increased risk of records being lost or compromised.
Storage standards for ongoing trial materials	Authorised access to records.

Criteria	Procedures
Recommended archive standards	All archive processes documented, approved and reviewed regularly.
Minimum archive standards	Archive processes documented.
Risk of using minimum archives standard	Processes not defined and followed.
Storage standards for ongoing trial materials	Working practice agreed.

Criteria	Auditing
Recommended archive standards	Internal quality audit program of all aspect of archive systems and procedures on a regular basis.
Minimum archive standards	Regular assessment of archive operation.
Risk of using minimum archives standard	Lack of process improvement reflecting changes in regulations or other factors.
Storage standards for ongoing trial materials	Records must be made available on request.

Criteria	Destruction of records
Recommended archive standards	Authorised, secure, confidential disposal of materials, governed by SOP. Appropriate records kept.
Minimum archive standards	Authorised, secure, confidential disposal of materials, governed by SOP. Appropriate records kept.
Risk of using minimum archives standard	N/A
Storage standards for ongoing trial materials	No destruction of records without authorisation.

www.ingramcontent.com/pod-product-compliance
Ingram Content Group UK Ltd.
Pitfield, Milton Keynes, MK11 3LW, UK
UKHW041836200726
13854UKWH00003BA/1165